WEIRD SEA CREATURES ™

THE JELLYFISH

Miriam J. Gross

The Rosen Publishing Group's
PowerKids Press ™
New York

For Elana

Published in 2006 by The Rosen Publishing Group, Inc.
29 East 21st Street, New York, NY 10010

First Edition

Editor: Daryl Heller
Book Design: Albert B. Hanner
Layout Design: Greg Tucker

Photo Credits: Cover © Royalty Free/Corbis; p. 5 © David B. Fleetham/SeaPics.com; pp. 6, 17 © Paul A. Sutherland/SeaPics.com; p. 9 © Mark Conlin/SeaPics.com; p. 10 © Jeffrey L. Rotman/Corbis; p. 13 © Gregory Ochocki/SeaPics.com; p. 14 © Eleanora de Sabata/SeaPics.com; p. 18 © Doug Perrine/SeaPics.com; p. 21 © Kevin Schafer/Corbis.

Library of Congress Cataloging-in-Publication Data

Gross, Miriam J.
  The jellyfish / Miriam J. Gross.
      p. cm. — (Weird sea creatures)
  Includes index.
  ISBN 1-4042-3192-7 (lib. bdg.)
  1. Jellyfishes—Juvenile literature. I. Title.

  QL377.S4G73 2006
  593.5'3—dc22
                                    2005000727

Manufactured in the United States of America

CONTENTS

NOT JUST A BLOB OF JELLY

Jellyfish are puffy, bell-shaped creatures with **tentacles** that trail below them. The nature of the jellyfish was long unknown to scientists. This was because these creatures were seen only when they got caught in nets or when their bodies washed up onshore. Since the jellyfish's body breaks apart when it is removed from the water, the jellyfish looked like a blob of jelly.

Today we know that jellyfish are among the world's most plentiful animals. There are more than 200 **species** that come in a range of colors and sizes. The largest is the lion's mane jelly, which can grow up to 8 feet (2.4 m) across, with tentacles measuring 100 feet (30 m) long. One of the smallest is the tiny Irukandjii jellyfish, which is only about ¾ of an inch (2 cm) wide. Most jellyfish have clear, colorless bodies, such as that of the moon jelly. Others have bright colors and patterns. The lion's mane jelly is orange, and the purple-striped sea jelly has purple stripes.

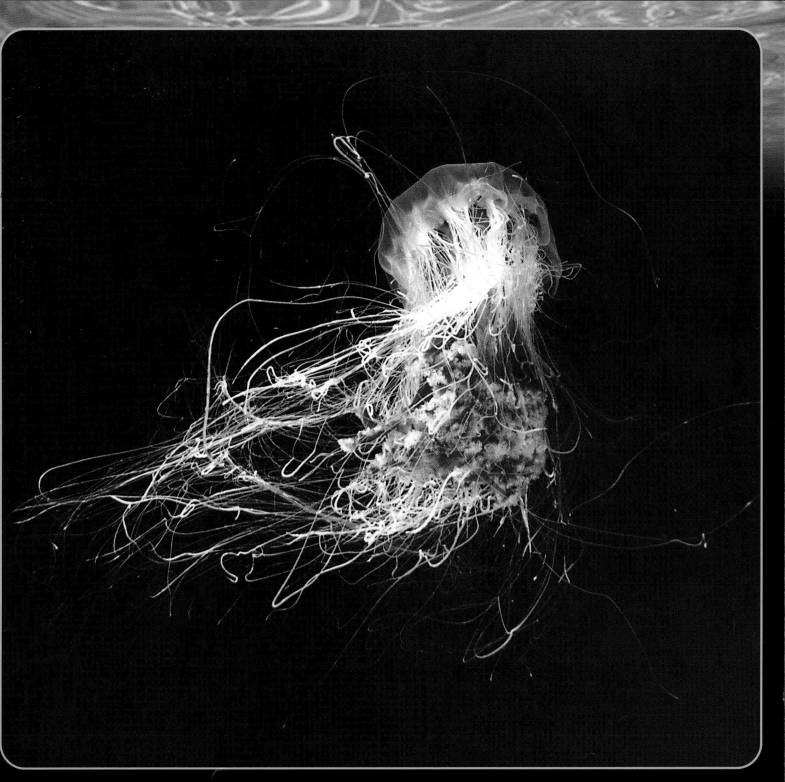

The lion's mane jellyfish has eight groups of tentacles. Each group has between 70 and 150 tentacles. This means that the lion's mane has between 560 and 1,200 tentacles in total. The scientific name for the lion's mane is Cyanea capillata.

This is a close-up of the tentacle of a sea wasp jellyfish. The rounded area on top is the cnidoblast. The cnidoblast holds the stinging cells, or nematocysts. The sting of a sea wasp, whose scientific name is Chironex fleckeri, can be deadly. The sea wasp is found in the Pacific Ocean off the coast of Australia.

A FAMILY OF STINGERS

Jellyfish are not really fish. They belong to a group of animals that scientists call cnidarians. **Sea anemones** and **coral** are also part of this group. All cnidarians (pronounced ny-DAR-ee-ens) have a soft, jellylike body, tentacles, and stinging cells, which are called nematocysts.

A nematocyst consists of a stinger inside a hollow tube. When something touches a jellyfish, a stinger fires from the tube and pierces, or sticks into, the victim's skin. The jellyfish then grows a new nematocyst in place of the one it has just fired. Scientists have found more than 30 different types of nematocysts. Some inject a poison into the victim's skin that paralyzes the victim. When a victim is paralyzed, he or she is not able to move. Others have tiny hooks, or sharp parts that stick onto the **prey**, holding it still so the jellyfish can eat it.

Stinging cells are found on the tentacles and on other parts of the jellyfish's body. The main purpose of the jellyfish's sting is to help it catch food. The painful sting can also **protect** the jellyfish from being eaten by other animals.

A SIMPLE ANIMAL

A jellyfish is a simple animal with no brain, no heart, and no **gills**. The **tissue** that makes up its body is so thin that **oxygen** can pass directly into it. The jellyfish has a mouth that is located in the center of the jellyfish's underside. Food passes into the mouth and is digested, or broken down into energy, in a hollow space in the middle of the jellyfish. Waste is then passed out of the mouth.

The jellyfish's body has two layers, or levels, of skin cells, with a thick layer of a jellylike matter in between. Most jellyfish are shaped like an umbrella. Stinging tentacles hang from the outside edge of this umbrella shape, which is sometimes called the bell. Some jellyfish also have a set of frilly arms around their mouth that is called the oral arms. The oral arms are used to move food into the mouth.

A jellyfish can swim slowly up or down by **pulsating** its bell. To do this the jellyfish pulls in the **muscles** around the sides of its body. This causes water to shoot out from under the bell, which pushes the jellyfish along.

This is a close-up of the oral arms that surround the mouth of a giant jellyfish. The scientific name for the giant jellyfish is Chrysaora achlyos. The giant jellyfish eats zooplankton, which are tiny animals. Another creature that eats zooplankton is the commensal crab. Because commensal crabs have a hard shell they are not hurt by the jellyfish's stingers. These crabs are eating zooplankton on the jellyfish's oral arms.

9

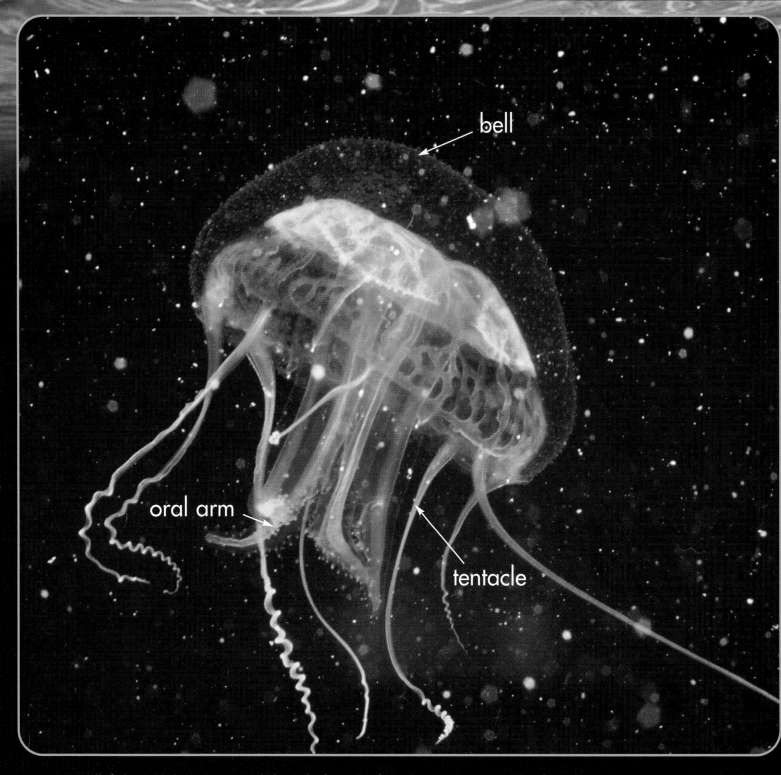

bell

oral arm

tentacle

Jellyfish were living on Earth long before the first dinosaurs were. The first dinosaurs were on Earth about 220 million years ago. Scientists have found fossils, or marks and remains left by jellyfish, that are more than 650 million years old.

LIFE WITHOUT A BRAIN

The jellyfish's body is made almost completely of water. This creature does not have organs that allow it to taste, smell, and hear like many other animals do. Organs are the parts of an animal's body that do a certain job. The jellyfish does not have a brain. It does not have a nervous system to carry messages from the brain to other parts of its body either. Instead it has touch **receptors** and special cells called chemoreceptors that carry out these tasks.

The touch receptors are located on the creature's tentacles and around its mouth. The touch receptors allow the jellyfish to sense movement in other animals and help it capture food. Instead of having a nose or a tongue, the jellyfish has special cells all over its body called chemoreceptors. These chemoreceptors allow the jellyfish to smell and taste.

Some jellyfish species can sense light. This helps them tell the sunlit surface of the ocean from the dark bottom. The light is sensed by special cells that are set on the bottom of the bell. These special cells are called ocelli.

JELLYFISH EGGS

The jellyfish's reproductive organs form in the lining of its stomach. Reproductive organs are the body parts inside an animal that allow it to make babies. In transparent, or clear, species, such as moon jellies, you can see these organs. They are the four circles near the center of the jellyfish's bell. To **mate**, the male lets reproductive cells out into the water through his mouth. The female takes in the reproductive cells and uses them to **fertilize** her eggs. She will hold the eggs in her oral arms. In some jellyfish species, the eggs are fertilized in the water when the reproductive cells join.

In many species the mother may carry thousands of fertilized eggs at a time. These eggs stay on the mother's oral arms until they grow into tiny, flat larvae. Larvae are in an early stage in an animal's life. The larvae have different forms than the adults they will become. Once jellyfish reach the larval stage, the mother lets go of the larvae. The tiny creatures then drift in the water currents until they find a hard surface, such as a rock, to which they can attach themselves.

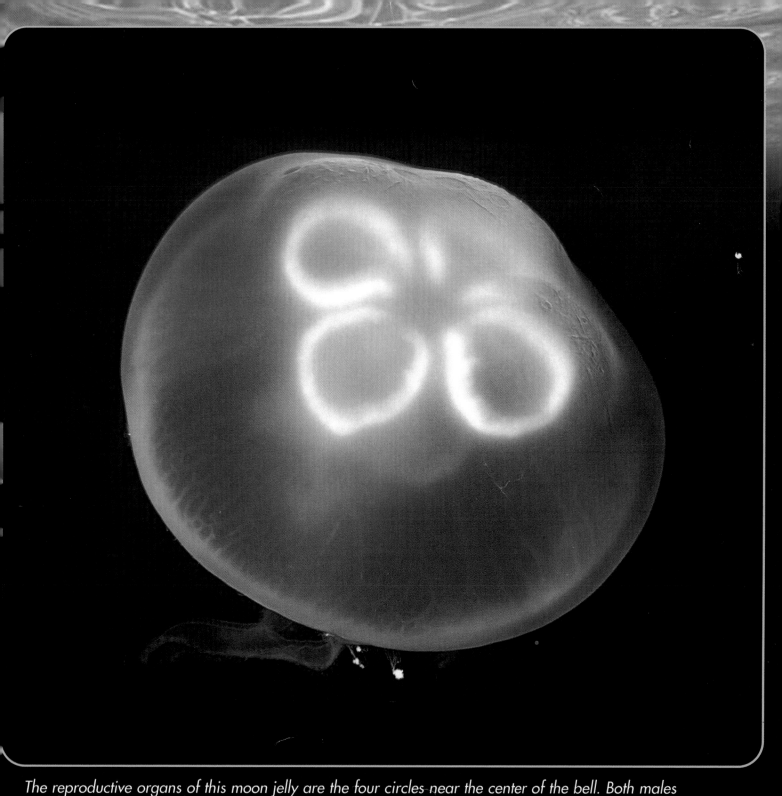

The reproductive organs of this moon jelly are the four circles near the center of the bell. Both males and females have these four circles. However, on the female the reproductive organs are yellowish. On the male they are purple. The scientific name for this moon jelly is Aurelia aurita.

13

The creature at the center of the photograph is a moon jelly in an early stage of its life. This moon jelly is an ephyra, or a jellyfish that is no longer a polyp, but not yet a medusa, or an adult jellyfish. This ephyra will grow into a medusa in a few weeks.

14

A Cycle of Change

Jellyfish usually live from three to six months. During this time their bodies go through many changes. The form the jellyfish takes at each stage changes so much that the jellyfish looks like a different animal. The ability to change in this way is called polymorphism.

Once the larva settles on a hard surface, the creature attaches itself to the surface and becomes a polyp. The polyp, which cannot move, looks like a plant. After several months **grooves** start to form on each polyp. Finally the grooves cut through the polyp, turning a single polyp into a stack of many tiny disks. These disks are about ⅛ inch (0.3 cm) across. Each disk will become a new jellyfish. One by one they break off from the stack and swim away. The young jellyfish are now called ephyrae. An ephyra is a small **version** of the adult. In a few weeks, the ephyra becomes a medusa, or an adult jellyfish. This is the final stage in their life cycle.

The adult form of the jellyfish is called a medusa because of the tentacles that surround the creature's round body. Scientists named this stage after Medusa, a figure in ancient Greek stories. Medusa had snakes for hair.

EATING EVERYTHING

Jellyfish will eat anything they can catch. Most of the animals they eat are smaller than they are. These animals include baby striped bass, young blue crabs, and **plankton**. Many jellyfish species eat other jellyfish. Sometimes a jellyfish can be seen through the clear body of the jellyfish that has just eaten it.

The jellyfish catches its prey with the tentacles that hang below its body. As prey swims into the tentacles, it sets off the jellyfish's stingers. In most species the sting paralyzes the prey, so the prey cannot move and is unable to feel anything. The jellyfish then moves the prey to its mouth with its oral arms, where it swallows the prey whole.

Some jellyfish, such as the upside-down jellyfish, produce food though **photosynthesis**. The surface of this jellyfish is sprinkled with brown spots. The spots are **algae** that live in the jellyfish's tissue and produce food for the jellyfish using the Sun's energy. To give the algae enough sunlight to make food, the upside-down jellyfish spends most of its life flipped upside down. In this way the algae in the jellyfish face the Sun.

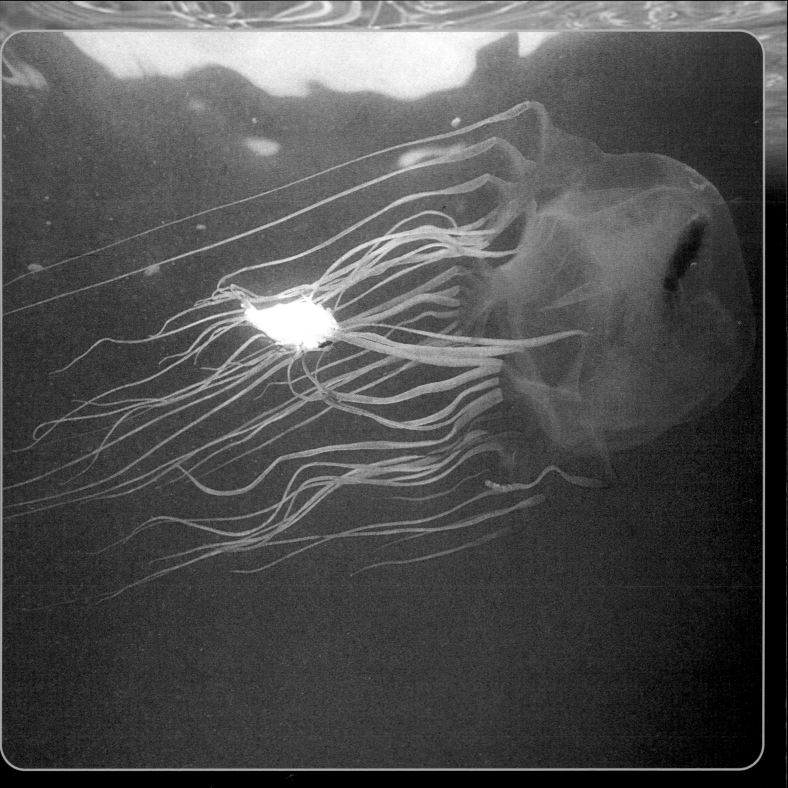

This sea wasp jellyfish has caught two different prey. Inside the creature's bell is a fish, which will soon be digested. Trapped in the sea wasp's tentacles is another fish.

This green sea turtle is eating a moon jelly in the Pacific Ocean near Hawaii. When green sea turtles are young, they are mostly carnivores and eat animals such as jellyfish. When sea turtles become adults, the food they eat changes. Adult sea turtles are mostly herbivores, which means they eat plants.

A Jellyfish Snack

Jellyfish are an important part of the ocean's food chain. They provide **nutrition** for many kinds of fish, including tuna, salmon, and ocean sunfish, as well as other jellyfish. Although the sting of a jellyfish bothers some of these animals, it does not hurt them enough to stop them from eating jellyfish. A sea turtle will continue eating a jellyfish even as the stingers make its eyes swell with pain.

Jellyfish bodies are mostly water, so some animals have a special way of eating only the nutritious part of the animal. Sunfish and leatherback turtles both have special teeth at the back of their throat. After swallowing a jellyfish, the animals will **regurgitate** it against these teeth. This action strains out the water and saves the **edible** tissue, which they then eat.

Sadly sea turtles often mistake plastic bags or balloons in the water for jellyfish. When they try to regurgitate them in the usual way, they end up choking to death.

JELLYFISH AND THEIR WORLD

People sometimes cause the jellyfish population to increase by dumping waste into the ocean. Waste such as fertilizer, which farmers use to make plants grow faster, causes an increase in the number of the algae on which jellyfish feed. When there is so much food to eat, jellyfish multiply quickly, until they outnumber their neighbors.

Jellyfish are found in oceans around the world and even in some freshwater lakes and rivers. They can live in the freezing cold waters of the Arctic and Antarctic or in warm tropical waters such as Australia's Great Barrier Reef. The tropics are the warm parts of the Earth where the weather is warm throughout the year.

Some species live in shallow, or not very deep, waters near the coast. Others live in deep water, down to 12,000 feet (3,658 m) below the surface. Most jellyfish spend all their time swimming or floating in the open ocean. They do not have regular homes.

Jellyfish help their **habitats** by providing food for many other animals. However, jellyfish can also harm their habitats. If there are too many jellyfish in one area they will eat up all the food that other animals might be eating.

These jellyfish were photographed at the Monterey Bay Aquarium in California. This aquarium has one of the world's largest year-round displays of jellyfish.

JELLYFISH AND PEOPLE

The stings of common jellyfish, such as the moon jelly, can be painful. However, the sting of the box jelly is so harmful that it could cause a person's heart to stop. Swimmers should avoid waters where many jellyfish live. Even dead jellyfish or **severed** tentacles can sting. As a result it can be harmful to touch any jellyfish that wash up on the shore. In some areas, such as Maryland's Chesapeake Bay, there are alerts, or notices, to warn swimmers of the presence of stinging jellyfish.

Although jellyfish can be harmful to people, they are also useful. In some parts of the world, including China, Japan, and Korea, people eat jellyfish. They are salted and dried, pickled, or soaked in salt water, and then steamed. Scientists are also studying jellyfish to see if the **chemicals** in their bodies can be used to treat cancer and other illnesses. The cannonball jellyfish, which lives in the Gulf of Mexico, may produce a **substance** that can help treat painful illnesses, such as rheumatoid arthritis. Not only do jellyfish help people, but also they are an important part of their ocean home.

GLOSSARY

algae (AL-jee) Plantlike living things without roots or stems that live in water.

chemicals (KEH-mih-kulz) Matter that can be mixed with other matter to cause changes.

coral (KOR-ul) Tiny sea animals.

edible (EH-deh-bul) Fit to be eaten.

fertilize (FUR-tuh-lyz) To put male cells inside female eggs to make babies.

gills (GILZ) Organs, or body parts, used for breathing.

grooves (GROOVZ) Dents or cuts in the surface of something.

habitats (HA-beh-tats) The surroundings where an animal or a plant naturally lives.

mate (MAYT) To join together to make babies.

muscles (MUH-sulz) Parts of the body under the skin that can be tightened or loosened to make the body move.

nutrition (noo-TRIH-shun) The food that living things need to live and to grow.

oxygen (OK-sih-jen) A gas with no color, taste, or odor that is necessary for animals to breathe.

photosynthesis (foh-toh-SIN-thuh-sus) The way in which green plants make their own food from sunlight, water, and a gas called carbon dioxide.

plankton (PLANK-ten) Plants and animals that drift with water currents.

prey (PRAY) An animal that is hunted by another animal for food.

protect (pruh-TEKT) To keep from harm.

pulsating (PUL-sayt-ing) Growing and contracting with strong, regular movements.

receptors (rih-SEP-terz) Special cells that take in messages.

regurgitate (ree-GUR-juh-tayt) To throw up partly eaten food.

sea anemones (SEE uh-NEH-muh-neez) Soft, brightly colored sea animals that look like flowers.

severed (SEH-vurd) Broken off.

species (SPEE-sheez) A single kind of living thing. All people are one species.

substance (SUB-stans) Any matter that takes up space.

tentacles (TEN-tih-kulz) Long, thin growths usually on the head or near the mouths of animals, used to touch, hold, or move.

tissue (TIH-shoo) Matter that forms the parts of living things.

version (VER-zhun) Something different from something else, or having a different form.

INDEX

WEB SITES

Due to the changing nature of Internet links, PowerKids Press has developed an online list of Web sites related to the subject of this book. This site is updated regularly. Please use this link to access the list:

www.powerkidslinks.com/wsc/jellyfish/